FIELD OF SWABBING

Khosrow M. Hadipour

To order additional copies of this book, contact:
Xlibris
1-888-795-4274
www.Xlibris.com
Orders@Xlibris.com

FIELD
OF
SWABBING

Khosrow M. Hadipour

The subject material contained herein is based on forty-one years of oil field experience working offshore and onshore in Mississippi, Louisiana, Texas, New Mexico, the Gulf of Mexico, and Venezuela.

It is not the purpose of this book to be used as final procedure and/or definite guidelines. This book is prepared to act as a basic reference to swab tools, equipment, and their application based on my experience only.

Definition of Swabbing in Oil and Gas Operations

Swabbing is the artificial pulling of a well fluid by mechanical tools.

Swabbing is the artificial removal of water, oil, and gas from a well, using mechanical tools, which is quicker and cheaper (*fluid* can refer to oil, water, and/or gas).

Components of Mechanical Swabbing Tools

- Steel wire line cable referred as swab cable, the swab line, or sand line
- Hydraulic oil saver
- Swab lubricator
- Swab mandrels
- Rubber or steel swab cups of various sizes

Artificial swabbing begins when an oil or gas well cease to flow naturally. Once an oil or gas well stops flowing fluid, a swab process will begin to remove heavy fluid out of the wellbore and to prolong the natural flow as long as economically practical.

Artificial swabbing will reduce the hydrostatic head and help a well to flow.

Hydrostatic head is defined as the pressure exerted by a column of fluid.

Hydrostatic pressure = $0.052 \times$ fluid density $\times$ fluid height.

Effective well swabbing begins when the reservoir pressure depletes down, preventing a well from flowing. When the reservoir pressure declines below the hydrostatic column of fluid in the wellbore, then the well will stop flowing.

Methods of Swabbing in Oil and Gas Operations

- Through tubing string swabbing requires a good tubing string, packer, and seating nipple.
- Through casing string swabbing requires fully open perforations and a good casing string.

Only through tubing swabbing are the swab tools made up or screwed at the end of $^9/_{16}$ in. wire line cable (swab line) and lowered in the hole to a predetermined depth to artificially pull fluid out of oil and gas wells.

Artificial Fluid-Lifting Methods in Oil and Gas Wells

- Artificial lifting fluid by mechanical swab tools
- Artificial lifting fluid by beam pumping and sucker rods
- Artificial lifting fluid using gas-lifting tools
- Artificial lift using electric submersible pump
- Artificial lift by jet pump and hydraulic lifts
- Artificial lift using plunger lift

This book will examine the most common swabbing method, which is swabbing by using mechanical swab tools.

Tubing Swabbing

Through tubing swabbing is a method of artificially removing oil, gas, and water fluid from a predetermined swabbing depth out of a wellbore.

Tubing swabbing is used on different stages in oil and gas operations:

- Swabbing before perforating an oil or gas well, which is referred as underbalance perforating
- Swabbing after perforating with the intention of pulling water and making the well flow naturally (referred to as bringing the well in)
- Swabbing after an acid treatment (to pull acid and solids out of the wellbore before fracturing treatment or running an artificial pump in the well)
- Swabbing the water block in oil and gas wells to remove the hydrostatic head to make the well flowing again

Tubing Swabbing Preparation

- Check location, and remove all hazardous objects.
- Check and record well pressure. Don't swab in any well with flowing pressure of more than 10 psi.
- Move in and rig up the swabbing unit
- Rig up a swab tank and flow line from the well to the rig tank. The swab tank must be located 150 ft. away from the wellhead, behind guy lines.

Swabbing High H₂S Wells

Hydrogen sulfide (H2S) is a very toxic element in nature, and dangerous to human health. H2S is heavier than air and smells like rotten eggs. H2S is referred to as a silent killer. Sour crude oil is the source of hydrogen sulfide gas in solution.

- Everyone on the well location site must wear breathing and protective monitoring devices.
- Sensor and alarm devices must be provided for everyone working on the well.
- Never climb or bend over a swab tank containing H_2S, acid or any harmful gases.
- Never enter any rig tank, or frac tanks, containing hazardous fumes (contained space permit).
- All the safety protective equipment must be used (Personal Protective Equipment (PPE) requirements).
- Use wind sock or flags when swabbing on H_2S wells.
- Remove all flow line chokes, checks, and restrictions (make sure the line is not blocked with any foreign objects, such as paraffin, mud, and/or solids).
- Check tools, especially sand line, rope socket, and oil-saver rubber to make sure they are in good standing.
- Do not allow anyone to stand or work close to the rope or rope drum while operating.
- Do not work on a well with H_2S, unless proper safety protection is provided.

Tubing Swab Procedures

Move in the swab unit and rig up on the well. Adjust blocks and swab lubricator hanging at the center of wellbore.

There are basically two types of swab units in the oil and gas field:

- *Single swab mast.* These units are designed to swab oil and gas only. The single swab mast is very quick to rig up and rig down on a well and is more cost effective.
- *Workover pulling units.* Workover rigs are designed to pull and run tubing string and conduct remedial well repairs, completion, fishing operations, and wellbore cleaning task. Some pulling units are equipped with "double drums" advantage and are capable of performing well swabbing as well as other types of well work.

Make sure the support legs of the swab unit are set on a hard surface around the cellar and that the unit is set sufficiently leveled. Use safety anchors to support the swab unit if necessary.

- Spot the swab tank behind guy lines 150 ft. from the swab unit.
- Lay the flow line from the wellhead to the swab tank.
- Remove all flow line restrictions.

Check all the sheaves, shafts, and pulleys to make sure of the following:

- They are not misaligned.
- Sheaves are greased and rotating smoothly and turning freely.
- There are no cracked, chipped, or badly worn sheaves.
- Sheaves and pulleys are free from dirt and paraffin buildup.

Start examining the tools very closely to avoid breakdown during the swab operation.

The swab line is often referred to as a swab line, swab rope, or sand line.

- Check the swab line and rope socket before you start swabbing to avoid losing the cable in the hole.
- Make sure there is enough swab line on the drum to swab a well as deep as required.
- Make sure the swab line and rope socket are in good condition before the start of swabbing.
- Change the rope socket after each deep well swabbing.
- Change the rope socket after a lengthy swab operation.
- Change the rope socket if there is a sign of a bad rope socket connection.
- Change the rope socket after any acid swabbing.

Making a Rope Socket

The rope socket is a very important connecting point where the swab line and tools are connected. Making a rope socket must be carried out with accuracy and skill.

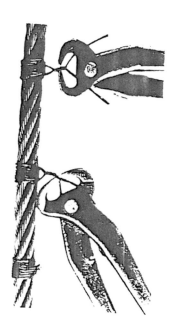

A) To make a rope socket, lay the wire rope flat on the ground and examine the cable from the hoist drum to the end of the rope where the new rope socket is to be made.

B) Make sure there are no kinks, twists, or other damages along the length of the rope. Always tie the rope securely with wire and wrap five loop knots around the rope strands to hold them in their original position above and below the space where the rope socket will be located.

C) Make three separate knots, 5 in. apart, and one knot above the distance where the rope socket will be located.

D) Saw out or cut the rope above the top knot.

E) Unwrap the top knot, separate wires, and cut out the core. Start to separate and broom out all the wires.

F) Wash and clean up the broom wires in kerosene, gasoline, naphtha, or verso, then dip the wires in an equal solution of weak acid (muriatic acid and water). Dry out the wires properly. Heat up the old socket to melt down old Babbitt using a pair of clamps. Pull and remove all wires out of socket. Clean up socket basket as best as possible. Cool off the tool.

G) Bunch wires together as closely as possible and force them into the socket basket.

H) Arrange the wires in the socket evenly using a screwdriver. Bend outside wires slightly to resemble a hook.

I) Preheat the basket at the socket using a pair of clamps to work with.

J) Start to melt zinc, Babbitt. The molten Babbitt must be heated to a correct temperature of 750°F and/or hot enough to flow freely.

Once the melted Babbitt is ready, pour the molten Babbitt into the socket and the wires in an upright position. Allow the hot rope socket to cool off by air. Do not cool with water.

Checking the Swab Lubricator and the Swab Tools

A) Lay down swab lubricator with swab tools, flat and off the ground.

B) Check the rope socket to make certain the wire rope and the rope socket are molded and engaged sufficiently and properly.

C) There must be no sign of looseness between the molded rope wires and the socket.

D) Check the rope socket to ensure the cable and rope socket is in good and safe condition.

E) Cut the cable and make a new rope socket if necessary to avoid losing the tools in the hole (a good rope socket will have the lead Babbitt right to the top of the fishing neck).

F) Make up the rope socket on swivel and sinker bars.

G) Check the swivel for flat or galled threads. Do not overtorque the connections.

H) If the rope sock is questionable, cut the cable and make a new rope socket.

I) Make up swab mandrel and swab cups (rubber swab cups or wire type swab cups).

J) Check and caliper gauge ring and no-go to make sure the tools are the correct size and will not go through and/or lodge through the seating nipple in the well.

Note: Ask the well owner/operator for correct downhole information. There are different sizes of seating nipples—API sized seating nipples, non-API seating nipples, X-nipples, XN-nipples, and seating shoes. Make sure to choose the correct size seating nipple.

Swab Mandrels and Knuckle Joints

Swab mandrels are designed specifically to carry the swab cups. These mandrels are designed and machined or casted with internal or external fluid bypass to allow the well fluid above and below the tool to pass through with ease.

Swab mandrels are made of high-strength steel or aluminum alloy for a variety of applications. Knuckle joints are used for the same purpose as swab mandrels.

KNUCKLE JOINT
WITH
MV CUPS

TA
WITH
RUF CUPS

STANDARD
WITH
J CUPS

UF
WITH
TUF CUPS

MACARONI
UF
WITH
UF CUPS

Today, knuckle joints have better, easier, and quicker application than other types of swab mandrels.

Knuckle joints or swivel joint mandrels are used more often by swab operators due to easiness in replacing the swab cups in a shorter period of time without having to lay the swab tools down.

Knuckle joints are designed to pass through dogleg and crooked holes more easily due to bend ability and flexibility at the knuckle and the joint.

Knuckle joints can be linked together in single, double, or triple if necessary.

Check the swab mandrels, especially the knuckle joints, and swivel joints mandrels to make sure the tools are not cracked, worn out, or bent.

If using other types of mandrels, such as ball and seat type or one piece mandrels with no-go nut, make sure the ball and seat and nut are in good running condition.

Caliper and gauge the no-go to make sure it will not go through the seating nipple. Any evidence of dent, rust, and trash between the ball and seat will stop the tools from holding fluid.

All the swab mandrels with a seating nut, acting as both no-go and the swab cup seat, must be checked properly. The pipe wrench scars or deep scratches at the surface of engagement of swab cups and mandrels will create leaks and fluid loss.

Check and caliper the no-go or gauge ring on the tools to make sure it will not lodge through seating nipple to prevent it from getting stuck and avoid unwanted fishing.

The Advantage and Limitations of Ball- and Seat-Type Mandrels

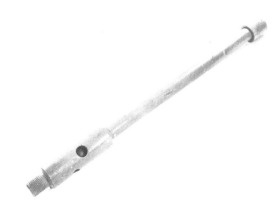

Passing fluid through the mandrel (internal bypass) will eliminate some of the swab cup damage due to fluid cut while the swab cups are hit with hydraulic force going down the hole. This method will bypass the fluid better than the mandrels with the external grooves and also reduces some of the impact fluid force when the swab tool hits the fluid level. The major problems that may limit the use of the ball and seat mandrels are sand, dirt, and solids, which may exist within the wellbore fluid.

If sand, trash, etc. land between the ball and seat, the swab load will be lost in the hole during the trip up the hole.

Damaged ball and seat as a result of corrosion, fluid cuts, etc. will also create similar problems. The ball and seat must be checked periodically.

Damaged seating nut due to the application of a pipe wrench and galled threads will often create problems.

Periodic inspections of matched ball and seat is necessary to prevent a dry run on this particular mandrel.

Slip-Type Mandrels

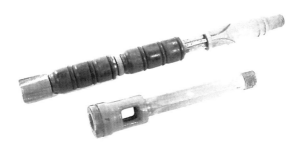

This type of mandrel is usually a single mandrel with a bottom screw-type nut acting as a no-go. Slip-type mandrels are externally grooved for fluid to bypass.

These mandrels are made with or without shear pins. They are made of either aluminum or steel bars.

To replace the cups, swab assembly should be laid down on the ground to unscrew and/or screw the mandrel bottom nut.

The Swivel-Type Mandrels

Swivel mandrels are similar to knuckle joints with a difference that swivel type may rotate more freely. The mandrels are also grooved externally, allowing the fluid to pass above the swab cups when lowering the swab tools through fluid.

It is easy to replace the swab cups in just a few seconds by pulling the joint section out without laying the swab tool on the ground.

During the engagement of the tools with swab cups, the mandrel joint must be turned ¼ in. around to lock the joint in position.

The Swab Bars

Swab bars are solid steel alloy, which are made of high strength material. Swab bars are available in length of 2, 3, 5, 10, and 20 ft. with variable sizes of ¾, ⅞, 1, 1⅛, 1½, and 1¾ in.

Since paraffin, scale, oil, water, and gas pressure are present in most oil and gas wells. Heavy swab bars are needed to overcome some of the down hole preventing factors.

The purpose of solid bars is to provide sufficient weight for the tools to get down and to support the swab tools below. The weight bars can be added or be reduced easily and quickly as needed.

The Swab Cups

Swab cups are the most important part of the mechanical swab tool.

There are two types of swab cups:

- Rubber-type swab cups
- Wire-type swab cups

Both types of swab cups are safe to use if applied properly. Check your swab cups, regardless of type and shape, whether new or used.

Swab cups with cuts, dents, scratches, torn-up rubber lips, bent sleeve, or bushing must not be run in the well (use swab cups with aluminum sleeves).

If choosing to use wire swab cups, file down the tips of wires from the outside in, to an angle.

The Tubular Jars

As mentioned in previous pages, the purpose of the tubular jars is to release possible stuck swab cups from tight spots, scales, or sand bridges, etc.

For better jarring action and sufficient act of performance, the tubular jars must be clean and without damage. Jars must function freely before put to use.

All elements such as paraffin, mud, cement, sand, and dirt must completely be removed from the tools. Existence of the abovementioned materials inside the tubular jars and around the bars will prevent the tool from satisfactory stroking.

Regular inspection on all the tools is recommended. Tubular jars ought to be located below the sinker bars or weight bars and above the swab mandrels and the rubber cups for an effective jarring action. Check the rod for bend or damage.

The Hydraulic Wireline Oil Saver

Safety and pollution prevention is dependent upon this tool.

Check to make sure the oil saver functions properly. Fill up the pump reservoir with hydraulic fluid, close the relief valve, and pump through the oil saver pump.

Make certain that the plunger inside of the oil saver tools is moving upward and downward. If this tool does not function properly, do not swab the well until the tools are repaired.

The Swab Lubricator—It Is Not Practical to Swab without Lubricator

All the wireline operations—such as logging, perforating, slick-line, and swabbing units—ought to operate through a lubricator to control well pressure and prevent blowout, pollution, and/or serious accidents.

The swab lubricator is made up on a joint of 2⅞ in. tubing with union connections. Make sure the lubricator is not split or does not leak at the connections before application.

In order to rig up or rig down the swab tools, the swab lubricator is designed with short section and longer sections of tubing joints with easy makeup hammer unions.

Parts of the swab lubrication must be tightened by using a hammer prior to the swab. Application of hammer force against the hammer union under pressure <u>may be risky</u>.

Swab lubricators must be kept clean and free from paraffin nest and solid buildup internally to avoid friction and prevent the swab tools from getting stuck. A dirty swab lubricator will cause cable damage and bird nesting.

Blowout Preventers

This tool is required for well control measure and to prevent fluid flow out of lubricator during a possible oil saver failure while swabbing.

The blowout preventer consists of housing with two rubber packing elements. The packing elements should be designed to fit and pack off tightly around the wire rope in service to prevent any leaks under unexpected well pressure, tool failure, or fire and to obtain a positive shutoff.

The preventer rubbers must be of correct size and in good working/operating condition at all times. The blowout preventer should not be used as a wire-rope wiper during a swab work.

This tool ought to be located above the master valve and just below the swab lubricator. All the swab mandrels must be equipped with a blowout preventer.

The blowout preventers must fully be in the open position during swab operation to avoid friction damage and bird nest problems while swabbing.

Pressure Gauges

A pressure gauge will serve as a monitoring device during swabbing trips. Well pressure may be detected during swab, and abnormal conditions, such as blocked-off flow line or pinched surface valve during swab operation, can be detected.

A pressure gauge is recommended to monitor well pressure coming out or going in the hole.

A pressure gauge is usually usually installed on the swab lubricator and above the BOPs (blowout preventers).

Remove pressure gauge when intending to use hammer in making up unions. After a complete and satisfactory checkup of the swab tool assembly, prepare the tools for the next swab run.

Here are useful suggestions before you start swabbing:

- Operator must ask and learn about the tubing string and the history of the well subject to swab.
- Learn about possible history of sand, paraffin, pipe scale, salt deposits, and formation build up. Learn about any and all the downhole restrictions, which might cause problems during swabbing (crooked holes, doglegs, weight on packer, tight spot, buckling, fishing, or hole in tubing string).

Consider all safety aspects as well as all the above information. Prepare to swab with care.

Prepare to swab through the tubing string:

- Lay down and check the flow line from the pumping tee at the rig floor to rig tank.
- Remove chokes, checks, and flow line restrictions from wellhead to the rig tank. Flow line must be fully opened from the well out to the swab rig tank.
- Check the oil saver assembly on the ground or on the rig floor. Prepare the oil saver split bushings and check the parts to make sure they are not broken or cracked.
- Put on a set of good brass bushings around the wire rope, then open up the oil saver rubber, and engage the oil saver rubber over the rope.
- Mount a set of brass bushings (bushings are facing the oil saver rubber).
- Check to make sure all the parts are set correctly and locked in place.
- Force the top bushing in and lock up the assembly with the U-bolt and key on top.
- Close the master valve and swab valve on the well.

- Pick up the entire swab tool assembly above the rig floor.
- Put one swab cup on gauge ring knuckle joint on the first swab trip.
- Mount the complete swab lubricator with a set of BOPs and new rubbers.
- Rig up the swab lubricator and prepare to flag the swab cable before going in the hole (swab cable may be called sand line or wire rope).

<u>Prepare to flag the sand line before swabbing:</u>

- Never swab fluid in any oil and gas well without swab <u>lubricator</u> and the <u>sand line flags</u>.
- There is no accurate depth correlation device in swabbing operation.
- Screw a crossover with union connection on top of the pumping tee above the Christmas tree for easy rig up and rig down of the lubricator.
- Make sure the swab valve and master valve is in the closed position when spacing.
- Pick up the swab lubricator with complete BOPs and connections
- Install tubing lubricator.

Start flagging the wire rope sand line.

Flagging a sand line is usually a duty of two roughnecks—one person to twist open the cable strands with a pipe wrench and one person to put on the nylon flags correctly and evenly in place.

The proper installation of the swab flags is important

- Close the master valve and swab valve on top of the tree.
- Pick up and install the swab lubricator.
- Slack off with swab tools and tag the closed swab valve above the wellhead with slight tension. (Avoid too much slack-off.)
- Pick up 3 ft. off the bottom into the swab lubricator.
- Mark the swab cable at the cable drum at the operator's eyesight level.
- Open the rope strands with a pipe wrench.
- Install the first set of nylon flags evenly at the operator's eyesight level—install the flags evenly about 12 in. apart.
- Avoid damage to the swab cable when opening strands with a pipe wrench. If wires break when twisting to open the strands, stop using the cable. The cable may not be good to swab with.
- Open the swab valve and master valve. Lower the swab tools in the hole until the first set of swab flags reaches the top of the swab lubricator.
- Install the second set of nylon flags as done above, using two sets of nylon flags spaced 8 in. apart.

- Continue going in the hole until the second set of flags reach the top of the swab lubricator.
- Install the third (3rd) and final set of nylon flags evenly at the operator's eyesight level, using two sets of nylon flags spaced 8 in. apart.
- Do not make heavy sets of flags. The heavy sets of nylon flags may ball up and get stuck at the oil saver rubbers, causing friction and birdcaging.
- Continue in the hole until the third or final swab flags reach the top of the swab lubricator. The operator may mark the swab line at the cable drum as a reference point.

Note that the nylon flags may deteriorate or be lost by pulling through the saver rubber. Renew the swab flags; at any time is recommended.

Prepare to swab the well:

- One person as experienced swab operator
- One person to watch swab flags coming out of the hole
- One person to measure fluid and catch the fluid samples
- One person to assist the swab operator, check tools, and change the swab cups

Note: the above mentioned duties is related to double-drum workover rigs. Single mast swab unit typically uses two people.

Now the tools are ready to swab fluid.

After flagging, you may pull out of the hole and check the swab flags and cable if necessary. Remove and check the swab lubricator and tool spacing once more.

Recheck the tools. If you have a swab valve below the lubricator, you may leave the master valve open. Give instructions to avoid closing the master valve on the swab cable by accident and losing the tools in the well.

Engage one swab cup on the swab mandrels with the face of the swab cup's lips looking up.

You may have a choice of choosing a swivel-type mandrel, knuckle joint mandrel, etc.

Apply one swab cup on the first run.

Turn the mandrel one-fourth turn for proper engagement. Now pick up and mount the lubricator onto the Christmas tree or onto the wellhead.

Bleed off pressure on the swab saver pump to allow the cable to go down the hole.

The swab cable may be slow going down the hole on the first run due to the new tight oil saver rubber and/or the wellhead's minor pressure.

If the wellhead pressure is 20 psi or more, stop swabbing. Bleed down wellhead pressure and start in the hole with caution.

Start going in the hole at a moderate speed while counting the drum wraps.

You may use a depth meter for more accurate measurements (and avoid accidents when installing and holding the depth meter on the sand line).

Do not run the tools down the hole too fast. Do not run the swab tools too far deep under the fluid.

If you're running the swab tools too deep under the fluid, it will be a difficult task to pull the tools out.

Watch your speed at all times, going in the hole or coming out of the well with fluid load. Expect gas pocket, oil, water, and/or tight spots at any depth.

The jarring counteraction between the swab tool tool and the fluid may cause serious damages to the swab cable, resulting in bad kinks and birdcaging or perhaps loss of the cable with tools in the hole.

Continue in the hole at a slow speed on the first run to feel for the wellbore's condition. Tag fluid and continue going down the well to 700 ft. below the fluid level.

Before pulling out of the hole, close the relief valve on the hydraulic pump. Pump up the plunger on the oil saver sufficient enough to wipe the swab cable and keep the oil, water, and gas from spraying out of the swab lubricator.

Slow down speed and watch fluid surge, approaching the surface.

Keep your eyes on the cable and expect the first flag to come out. As soon as the first set of flags appear on top of the swab lubricator, slow down and expect the second and the third set of flags.

When the last set of flags appear on top of the tubing lubricator, slow down and pull the cable slowly until the flags reach the expected spot at the hoist drum. The swab tools should completely be inside the swab lubricator and clear from the wellhead, master valve, and swab valve.

Always close the master valve slowly. Do not shut the master valve with the swab tool below the master valve (the master valve will cut and shear off the swab cable).

Note: It may be difficult to know if a gate valve is open or closed by looking (do not cheat on opening or closing a gate valve; always count the turns).

Close the well shut before removing the swab lubricator to check the tools, change cups, or rig down.

Continue going back in the hole to swab depth, which you may think is safe and sufficient load weight for the swab rope. Start pulling fluid load with a reasonably moderate swab speed per minute.

Continue with swabbing while checking tools, and measure swab fluid.

Check to make sure valves are not pinched at the surface. Start pulling at moderate swabbing speed, allowing the swab cup/cups to relax and adjust to the tubing condition and to avoid losing fluid coming out of the hole. Check swab cups for damage regularly to avoid losing rubbers in the hole. Report lost rubbers and/or any objects in the well.

Inspection of swab cups, the oil saver rubber and swab tool are necessary:

- P.O.O.H. (pull-out-of-hole) with swab tools into the lubricator.
- Close the well completely shut (make this a habit; do not leave well open).
- Break the tool at the union above the wellhead.
- Replace the cups, if necessary, and check the rest of the swab assembly.

When the inspection of tools is complete and you are ready to go back into the hole:

- rig up the swab lubricator with tools on the wellhead;
- open up the swab valve and master valve slowly, and check the well pressure;
- open up the relief valve on the hydraulic pump pressure;

- start going into the hole with caution, and expect gas pockets, tight spots, and water hammering while going down the tubing string;
- continue this procedure until you swab down the well satisfactorily;
- watch and monitor fluid rise on each swab run to prevent the swab tools from running away (operator has to retreat the swab tools) from high well pressure;
- when the well starts flowing, pull up into the swab lubricator. Wait, watch, and allow the well to flow and unload fluid into the rig tank; and
- monitor flowing pressure.

If the well starts flowing satisfactorily, then put the well on the production line.

- Close the crown valve below the lubricator only and rig down the swab unit.
- Turn the well over to the operator.
- Put all the wellhead equipment back on, just the way they were.
- Clean up the location if needed (pick up all used swab cups and trash).
- If further swabbing is requested, do not leave the tools rigged up during the night.

Present your daily report after the job properly and in a professional manner:

- Start swabbing time
- Stop swabbing time
- Depth at which you hit fluid on each swab run
- Depth you pulled fluid from on each swab run
- Fluid rise on each swab runs
- Number of swab runs at the end of day
- The last swab depth
- Total volume of fluid recovered at the end of the day
- Percentage of oil, water, and solids by taking swab samples (basic sediment and water (BS&W))

If repeating swabbing operation is needed, continue with swabbing operation as before:

- Check and keep flow line and valves fully open. Remove all the chokes before swabbing.
- Engage the swab cups on the swab mandrels with the face of the swab cup's lips looking up.
- You may choose the swivel-type mandrel, knuckle joints mandrel, etc.
- You may use one or multiple swab cups at the same time.
- Turn the mandrel one-fourth turn for proper engagement.
- Prepare the oil saver split bushings, and check the parts to make sure they are not broken or cracked before you start swabbing.
- Continue to swab fluid down tubing string—pulling gas, oil, and water out of wellbore.

Travel in and out with reasonable speed. Speed will damage swab cups and saver rubber. If losing fluid on trips, check swab cups. Check swab cup for size and look for cuts.

Going down the hole fast may hit gas pockets or tight spots. This may cause kinks on sand line and open the cable for bird nest (will discuss bird nest later).

When you start pulling fluid out, increase pressure on rubber saver enough to avoid the spray of salt water, oil, and gas pollution coming out of the lubricator. Avoid too much pump pressure on the rubber saver to prevent damages inside the rubber saver, causing fluid leaks.

Lubricate the tool more often to avoid wire flagging.

Once the fluid reaches the surface, slow down further to avoid fluid splash due to gas and oil pressure. Catch fluid samples. Record number of trips and volume of fluid pulled on each run.

Keep a good swab record:

- Use two swab cups on the gauge ring knuckle joint mandrel if necessary.
- Check swab cups and cable more often to avoid losing rubber pieces in the hole and to detect broken strands and wires.
- Check the swab cups for correct size (undersized cups versus correct size cups).
- Watch for the nylon flags coming out of the hole. Missing flags will result in losing the cable and may cause an accident (get help from the rig crew to watch for the flags while coming out of the hole).
- Continue swabbing as deep as necessary. <u>Do not swab to the seating nipple.</u>

- When the well starts flowing oil and gas, stop swabbing.
- Close the swab valve and allow the well fluid to flow down the production line.
- Rig down the lubricator and install the pressure gauge on top of the Christmas tree.
- Clean up the location and move off the location.

Advantages and limitations of through tubing swabbing:

- It can successfully swab a well as deep as 200 ft. above the seating nipple.
- It must make several swab trips to pull the required volume of fluid.
- Floating sand and mud may migrate up the hole as far as the wellhead.
- It needs an API seating nipple above the packer before one can start swabbing.
- The tubing string and packer need to be spaced 200 ft. above the open perforations.
- One must make sure the swab tool no-go will not go through or lodge into the seating nipple.
- The tubing string must be clean and in good condition to avoid getting stuck.

Casing Swabbing

The principle of casing swabbing is to use tubing string and casing swab tools without using swab line (no swab cable is used in casing swabbing operation).

Casing swabbing is used more often in saltwater disposal and injection wells to pull large volumes of fluid in one trip to clean up perforation tunnels.

Casing swab is used to pull sand, mud, and contaminants out of perforations by pulling swab tools harder and to create a hard vacuum on the casing in order to pull contaminated sand and mud out of open perforations. Cleaning the perforations will cause the formation to take more salt water with lower injection pressure.

The causes of saltwater disposal wells not taking water are the following:

A) Tight formation: contains clay, shale, reservoir pressure, etc.

B) Shot density and not enough open holes, scaled-up perforations

C) Near wellbore skin damage

D) Injecting dirty fluid with contaminated solids, oil, sludge, chemicals

E) Hole in casing: below the isolation packer and above the open perforations

F) Not enough rathole below the open perforations

G) Other mechanical problems

There are several remedial solutions to improve saltwater disposal:

A) Acidizing using compatible acids and chemicals

B) Application of coiled tubing to clean up contaminants

C) Through tubing swabbing to pull solids out of perforations

D) Through casing swabbing to remove contaminated sand and solids

E) Adding perforation for more disposal capacity

F) Abandon the zone and plug back higher to a better formation.

Salt water disposal well is the <u>single most important well</u> in the field to consider.

If you cannot dispose produced water, you may not be able to produce oil and gas. You can not drink it, sell it, spill it or haul it.

Often, it is necessary to remove mud, sludge, sand, and silt from the wellbore to keep perforation tunnels fully open in order to dispose water with higher rate and lower pressure.

There are several methods to clean up saltwater injection or saltwater disposal wells.

Application of Coiled Tubing

Coiled tubing is used to wash and clean up low-pressure, saltwater disposal wells.

Coiled tubing is the faster and more effective method to wash down and clean up the wellbore in one day with positive results if done correctly.

Water, foam, soap, and nitrogen are used to wash and clean up sand, scale, and solids out of low-pressure bottom-hole wellbores.

The major disadvantage of coiled tubing is the damaging of the internal plastic-coated tubing string.

If using coiled tubing to clean up the wellbore, make sure

    A) the tubing string is in good condition and without holes,

    B) the packer is seated and holding to avoid communication with annulus,

    C) there are no doglegs or bad spots in the casing string,

    D) there are no parted tubing or large holes across the open perforations,

    E) the mud motor is used to drill out hard scales in the tubing string.

Casing Swabbing

One must use a workover unit (pulling unit) to conduct casing swabbing.

Casing swabbing technique is an effective method to remove contaminated sand and solids out of saltwater disposal well. Casing swabbing may be effective in some wells and less effective on others.

Be cautious when using casing swabbing. Make sure the perforations are fully open before running casing swab tools to avoid breaking down the squeeze hole above the open perforation and/or collapsing the casing string.

Perforations must stay opened. Repeated casing swabbing may pull large quantities of formation sand and solids from behind the casing string and may create casing damage due to large cavity and creating voids.

Large fluid volume can be pulled on each swab trip.

Casing Swab Procedure

Rig up the service rig (a pulling unit). Pull the production tubing and packer out of the wellbore. Lay down the isolation packer. Rig up the tubing testers to drift and test tubing string to ensure no holes in the tubing.

Spot a rig pump and tank to swab wellbore fluid and to circulate the well.

Remove all the check valves and choke out of the return line line from the wellhead to the rig tank.

Install a fully opened, large return pipe from the casing outlet onto the circulating rig tank.

The return line must be as large as possible to prevent any back pressure while unloading the swabbed fluid.

Make up the casing swab tool at the end of one joint tubing or tubing stand and trip in the hole on the good standing tubing string.

- Check to ensure swab cups are the correct size.
- Ensure the swab cup is not damaged.
- Check to ensure swab tool is dressed correctly.

Start going in the hole with the swab tools slowly. The swab cups may be tight while going into the casing string due to light tubing weight. Pushing swab cups down the casing using pipe wrenches and blocks is normal procedure.

Continue going into the hole while watching the weight indicator.

When the swab cups hit/tags (meets) the fluid level in the casing string, the blocks and weight indicator will indicate less weight (tubing string will float due to fluid buoyancy).

Stop and make notes for tagging fluid level depth.

Continue going into the hole 1,400 ft. deeper below the top of the fluid level in the casing string.

Pick up and make three pipe strokes (60 ft. long strokes) with swab tools. You will notice a heavy weight gain (the weight indicator will show dynamic load consisting of the weight of the tubing, the fluid in the annulus, and the swab cups' friction).

Start out of the hole with the tubing as recommended by the swab operator (you may pull the tubing hard while coming out of the hole to suck the formation of sand and mud as much as possible into the wellbore).

Continue pulling the tubing out of the hole while unloading the wellbore fluid. Continue out of the hole and slow down below the wellhead. Pull and lay down the swab tools.

Check and replace swab cups if necessary (often the bottom swab cup will be damaged as the result of fluid load and hard vacuum below the swab cups).

If additional swab trip is needed, the open perforation must be checked to make sure all the perforations are fully open.

Trip in the hole with a mule shoe or rock bit on the tubing string. Watch for possible floating sand going into the hole. Stop and circulate the wellbore every 800 ft. with clean, fresh water.

Make sure to keep a good pipe tally.

Continue in the hole to the top of the sand bridge, if any. Wash and clean up the wellbore to plug back depth below the open perforations. Pull-out-of-hole (P.O.O.H) and lay down the mule shoe or rock bit.

Repeat the same casing swab procedure as many times as necessary to obtain satisfactory swabbing results.

Never swab the casing with with covered perforation (perforations must stay open when using casing swab tools). Swabbing with covered perforation may cause damage to the casing.

The casing swabbing tool consists of the following parts:
A) Short mule shoe
B) Cage with loaded spring and ball and seat
C) Shear pin sub for circulation, if necessary
D) Swab mandrel
E) Two (2) wire-type or rubber-type casing swab cups looking up
F) Spacer rings and lock ring
G) Tubing coupling of different sizes

Downhole Corrosion Attacks to Swab Wire Rope and Tubing String

Corrosion is defined as the loss of identity and the change toward the metallic natural state due to the reaction of metal with atmospheric moisture.

H2S, acid and organic gases may be the source of a corrosion attack.

Corrosion may be defined as the deterioration of a substance due to a reaction with with surrounding environments.

Swab line and tools are always the target of corrosion attack. This corrosion attack may vary with respect to the length of time, place, and well conditions.

Problems such as chemical attack to the wire rope, both internally and externally, as the result of organic acids (hydrogen sulfide [H_2S], CO_2, and other minerals) must be considered.

Stress corrosion is the result of heavy loads, tension. Corrosion may create rapid wire breaks, causing wire flags along the length of the swab rope.

The tubing flakes and internal blisters are the result of corrosion and rust and may often create swabbing problems (pipe scale).

Swabbing Problems

Swabbing problems may vary from well to well.

The source of problems during a swab job are two types:

- The wellbore problems
- The operator's negligence

The Well Problems

Paraffin buildup is due to a change in physical and chemical equilibrium of crude oil flowing up the wellbore.

The aggravating challenge of paraffin solutions in oil-producing wells, especially in the winter season, is time-consuming and costly to oil companies.

Paraffin is the result of change of physical state of hydrocarbon oil. The accumulation or buildup of paraffin inside tubular, downhole, and surface flow lines may create difficulties.

Depending upon the severity of the paraffin buildup, it is often impossible to get down with the swab tools.

Do not jar down with the intention of pushing the swab tool down the hole. Jarring on paraffin solids not only prevents the tools from going down but also may pack off paraffin content further down the hole and plug off the tubing string.

Crude oil paraffin is similar to honey wax, brown or soft-greenish with high viscosity. Paraffin is in solution in the hydrocarbon liquid originally.

Due to the well's bottom-hole temperature (BHT), the liquid wax flows up the hole with oil, gas, and/or water flow and gradually reforms and solidifies from its liquid state to a greasy substance and to a solid form like dry, hard mud.

The change in the temperature of oil and the migration of cold free gas will cause the paraffin wax to stick and build up in the tubing string as well as inside the casing string.

The paraffin buildup will continue on until it reaches a point where the paraffin completely plugs up the pipe. Some wells are more severely affected than others.

Paraffin may deposit from the surface to as deep as 5,000 ft., depending on the bottom-hole temperature of the reservoir. The higher the bottom-hole temperature, the less there will be paraffin deposits. The lower the bottom-hole temperature, the higher the chance of paraffin buildup.

Paraffin deposits also depend on the gravity of crude oil with high API gravity. Reservoirs producing cold oil produce greasy greenish paraffin more rapidly and must be cut very often to avoid plugging the tubing string.

Paraffin deposits vary from one well to another or from one reservoir to another reservoir in the same wellbore.

Some wells produce hot fluid, and others produce cold fluid. Wells producing hot fluids have a tendency to form less paraffin wax, while the cold fluid may form wax at shallower depths more rapidly and form paraffin wax when it comes out of solution.

Circulating oil in the rig tank during the cold winter will clearly show a rapid reform of thick paraffin wax buildup. Paraffin buildup is a change of the physical condition of crude oil from Newtonian to non-Newtonian fluid or pour point.

Melting point of paraffin ranges from 120°F to 150°F. At a temperature below the melting point, the chance of paraffin deposit increases.

These are the causes of paraffin buildup:

- Change in reservoir temperature
- Reservoir and surface flowing pressure

This is for remedial solution and paraffin control:

- Hot oiling at 200°F or over
- Steam application 250°F
- Removing paraffin by mechanical tools (wire line broach, scratchers, paraffin knife)
- Application of effective chemicals

Formation of Sand and Solids

Another major problem facing the swabbing operation is the existence of sand and solids.

Floating Sand

All the wells may produce some formation sand. None consolidated formations of Gulf Coast area will produce more formation sand than tight formation reservoirs.

The fluid subject to swab may contain various types of sand (BS&W)

The source of wellbore sand may be hydraulic fracturing sand or formation sand.

Formation sand grains vary in size. Some sand grains are coarse with a special color and texture. Some sand grains are very fine that they can stay suspended in the well fluid for a period of time before settlement.

These types of sand migrate from a producing well where the formation is unconsolidated or loose. This sort of sand has a tendency to pass the sand screens easily.

Sand and solids flow with oil, water, and/or gas into the wellbore (up into the casing and the tubing string). Sand may create problems during a swabbing process.

If the wellbore has been shut down for various reasons, sand deposit my plug up the tubing string in form of bridging and will hold high wellbore pressure from the bottom.

Sand bridge often causes serious downhole problems, especially on the rod pumps, gas lift wells, electric submersible pumps, and production equipment.

Swabbing on sandy wells is not a recommended practice.

Scale and Salt Deposits

Formation deposit is a fine compound driven from calcium, barium, calcium sulfate, and barium sulfate; Norms (naturally occurring radioactive material) are found in oil and gas wells.

Heavy scale deposits will make it very difficult to get down with swab tools. Attempting to swab may result may result in getting stuck and losing the swab tools in the hole. To swab in a well with scale deposits, run in the hole with only one rubber cup to 800 ft. under fluid carefully and pull out of the hole with fluid slowly.

Swab tools may partially hang up and release, which will indicate the trouble of scale deposits down the hole. Often the cups will be damaged and will not pull fluid out. Do not run steel swab cups in the hole by any means. All soft rubber cups are safer to swab with in scaled tubing string.

There are several types of scale deposits in oil or gas wells to deal with: calcium carbonate, calcium sulfate, barium sulfate, and iron sulfate.

Like paraffin, the scale is in solution in original form. As water and oil migrate to the surface, the fluid the fluid undergoes pressure change, which causes deposits of various scales. Scale deposits can also occur due to introduction to air and mixture of incompatible water and completion fluid from one well to others or introducing surface air into the wellbore.

Scale removal technique in oil and gas operations:

- Drilling—cut scale deposits using drill bit, and scrapers
- Wire line broaching—used for cutting or breaking hard formation deposit
- Chemical—some scale deposits are acid soluble and others insoluble.

Salt (sodium chloride [NaCl]) deposits can be found in oil and gas wells. Salt deposits will occur due to change of temperature, disturbing the solution equilibrium. Salt will form in oil and gas wells just like any other scale or paraffin deposits.

There is salt in the solution in the reservoir, and as fluid flows in and up the casing, the gas comes out of the solution, causing the oil and salt water to cool off and deposit crystal salt in the tubing and casing strings. Salt deposit in tubular goods is a very hard task to face.

Constant care is necessary to avoid solid salt bridges inside the tubing string. Salt may form inside tubing of several hundred feet. It is very difficult to swab through tubing with salt buildup.

Remedial workover in salt content well are often difficult due to high-pressure gas traps below the salt bridges.

Mechanical cut and pumping is very difficult in a prolonged salt deposits well. The wellbore must be treated with preferred hot fresh water and chemicals to prevent both salt deposits and paraffin buildup (Columbus, Texas, and New Mexico fields).

Fresh water steam is an effective solution in removing salt and paraffin deposits.

Buckling and Tight Spots in tubing string.

It is a difficult task to swab in a tubing string with tight spots and doglegs. Tubing buckling and tight spots in tubing occur as the results of:

- tubing handling-loading and unloading tubular (pipe) during transportation;
- tubing buckling, which results from too much slack-off or weights on isolation packer;
- dropping tubing in the hole
- jarring and slacking off on the tubing string during fishing; and
- sudden shift in casing due to parted casing and known force.

The arc of the tubing buckling depends upon the space between the casing and tubing annulus. The larger the casing, the sharper the tubing buckling (too much slack-off on $2^1/_{16}$ in. and 2⅜ in. tubing string inside 7 in. or larger casing string). The larger the casing, the sharper will be the tubing buckling.

Tubing buckling and doglegs will often prevent the swab tools and wire line equipment from getting through. Buckling and doglegs at a tubing collar will cause tearing off the swab cups and may result in dry swab runs.

Sharp dents may occur as the results of work-over, pulling unit slips, and tongues while tripping the tubing string in and out of the hole. It is difficult to pass through or swab through tight spots without getting stuck.

The causes of tight spots in the tubing string could be:

- manufacturer defect (uneven pipe drift and oval shape tubing)
- compression on the pipe and packer or anchor
- damaged tubing as the results of jarring, torque, and buckling
- damaged tubing during fishing jobs and snubbing work
- tight spot due to hard scale deposits
- collapsed tubing due to differential pressure and wall thickness
- collapsed tubing due to packer movements during high-fracture treatment pressure

Swab Operator's Negligence and Tool Failures

These are the causes of losing swab tools in the hole:

1) *Backing off swab tools.* This can happen if the swab operator fails to check and retighten the joints. Parts such as sinker bars, swab mandrels, tubular jars, and rope socket connections may unscrew due to many reasons. Regular checking on all the break points is necessary to avoid loss of the tools in the hole.

2) *Pulling speed.* The most adequate and safe swab speed is totally dependent upon the well condition, experience, type of fluid in the hole, the condition of swab tools, and sand line. However, field data indicates that the safest swab runs and speed of pull and runs may be suggested as follows:

   a) Speed on the first run down the hole must not exceed 200 ft./min. This will prepare and familiarize the swab operator with the well condition and locate the top of the fluid column in the wellbore.

   b) Succeeding runs thereafter should not exceed more than 300 ft./min. in the tubing without fluid, and when reaching close to the fluid column in the pipe, speed must be reduced to prevent a fluid pound, which may result in a kink and birdcaging in the wire rope or possible loss of swab assembly down the hole.

   c) While passing through in fluid, speed should not exceed more than 200 ft./min. Do not push the tool down the hole in order to gain speed. Fluid buoyancy force will tend to slow down the tool, and the cable will spin like a spring inside the pipe, causing the rope damage and twist.

Always go down while you have some tension on the rope rather than when there's too much slack. The maximum pull of 300 ft./min. is recommended while tripping out of the wellbore. Pull slowly the first 200 ft., let the fluid roll, and then bring up your speed gradually.

Since the condition of most perforated wells are hard to predict during a swabbing process, be prepared for unexpected fluid rise and well pressure. Be prepared in case of hitting a gas pocket on the next swab runs.

Continuous rising of fluid of several hundred feet per minute indicates that the well pressure may overcome the swabbing process. Be prepared to handle unexpected flowing problems.

If the well pressure tends to push the tools up the hole and stops them from going in the hole, pull the tool inside the lubricator and watch the well pressure closely. Make sure your tool is completely cleared from the wellhead in order to shut the well in.

Do not run your swab tool in the hole under a well pressure of higher than 10 psi flowing pressure. Attempting to get down the hole under pressure may result in throwing a bad kink or birdcaging and may damage the sand line, seriously making it difficult to pull the tools out.

Swabbing in new wells, which may have not been perforated yet, is safer. In those wells, there is no gas or oil in the hole. Even though the absence of gas, oil, etc. may be assured, a complete swab lubricator with the swab tool assembly is required.

When you swab in a well without perforations, the fluid level should not either rise or drop lower than the swabbed depth. Possible rise of fluid can be due to a packer leak, casing hole, or tubing leak that permits the liquid from the bottom or backside into the tubing, tending to equalize U tube. If the packer gives up or leaks, casing pressure usually goes on vacuum during a swabbing run. Tubing fluid level will be increased on each swab run.

If such a problem occurs during a swab job, stop swabbing and test the packer while leaving the tubing side fully open. Pressure up on annulus at 100 psi during swabbing, just to monitor any communications with casing.

During swabbing, usually the tubing will go on a vacuum force immediately after the swab cups are pulled up the hole. This is due to the air volume tending to replace the space of the pulled fluid in the pipe.

The vacuum would be diminished in a short period of time. In perforated wells, this condition may also be expected for a short period of time, even though well gas pressure may be present in the wellbore.

These are the suggested safe swabbing speeds:

1) Speed to run down the hole with no fluid

<u>Minimum</u>	<u>Moderate</u>	<u>Maximum</u>
200 ft./min.	250 ft./min.	300 ft./min.

2) Speed of running the tool in the fluid

<u>Minimum</u>	<u>Moderate</u>	<u>Maximum</u>
200 ft./min.	200 ft./min.	250 ft./min.

3) Speed of pulling out of the hole with fluid load

<u>Minimum</u>	<u>Moderate</u>	Maximum
250 ft./min.	300 ft./min.	400 ft./min.

Losing Swab Flags

Losing swab flags will cause confusion and hesitation for the swab operator. If the swab operator does not know that possibly he has lost his flags in the hole, he will probably run the tools into the swab lubricator by accident and lose the tools in the hole.

Always make more than one set of flags and know the spot at which your sand line wraps will be ended. Mark that particular spot on the hoist drum. This will be a solution in case of an emergency. The cause of losing the flags is mostly due to the friction between the oil saver and the bushings and perhaps the looseness of wire strands where the swab flags are engaged.

Load Problem

A load problem on a swab tool is mostly due to the operator's knowledge, experience, and often underestimating the swab depth. Use the depth meter whenever possible.

Pulling heavy loads will stress the wire rope and reduce the normal useful life of swab line. The size of fluid load and pulling depth depends upon the experience of an operator, weight of the fluid, size of the tubing, and physical condition of the swab tools.

Do not pull a load with more than seventy percent (70%) of tensile strength of your new swab line (break point).

If someone asks you to swab the well down to 8,000 ft., for instance, he does not mean pull 8,000 ft. of fluid in one run (this has happened before), so make sure you pay attention to what you are doing.

Pulling 8,000 ft. of salt water in one run is an impossible task to face. Pulling the rope off the rope socket or breaking the sand line will be expected. If the tools become stuck, do not work the cable under the wellhead pressure. This will cause the packing element to fail and cause well control issues.

For purposes of safety and security, do not pull more than 800 ft. of fluid on the first run of any job. This will give you an idea of what might be the well condition and what to expect on the next run.

Calculating the hydrostatic of fluid:

HP = 0.052 x depth x density of fluid

Swab fluid = 8.9 lb./gal.

Pulling 1,000 ft. of water

$0.052 \times 8.9 \times 1,000$ ft. = 462.80 lb.

Approximate Weight of Swab Load of Spent Hydrochloric Acid (HCl)

Percentage of Acid	Weight (bbl) of Acid in Pounds	Weight (bbl) of Acid in Kilograms
1	351	159.2
2	353	160.1
3	355	161.0
4	358	162.0
5	359	163.0
6	360	163.3
7	362	164.2
8	364	165.0
9	366	166.0
10	367	166.4
11	369	167.3
12	371	168.2
13	372	168.7
14	374	169.6
15	376	170.5
16	378	171.4

Wash down swab line with freshwater after swabbing acid. Acid or spent acid will penetrate and cause corrosion and permanent damages to the cable.

Swabbing Load Analysis

Possible loss of swab load may be as the result of the following:

A) *Torn-up swab cups.* Damaged or torn-up swab cups are due to several factors:

    1)   Dogleg and crooked hole

    2)   Heavy loads

    3)   Sand cuts

    4)   Friction and normal wear and tear

    5)   Manufacturer defects and mishandling or storing

    6)   Sharp objects such as tubing collars and heavy scale deposits

    7)   Collapsed swab cups.

    8)   Pulling speed and friction.

B) *Damaged seat on the swab mandrels.* A damaged swab seat mandrel will permit the fluid to leak through the swab cups. Dents, cuts, and nicks can occur due to throwing the tools inside the toolbox or normal dropping of other objects on them. A rusted swab mandrel and abnormal ring resulting from continuous use will certainly create loss of load.

C) *Damaged swab cup sleeve at the bottom of the swab cup.* Out of the roundness of the swab cup sleeve, especially the seat, this will allow the fluid to bypass down the hole. Loss of fluid will also result when the cup bushing or sleeve of the swab cup is exposed from the bottom. This will cause a poor seal-off contact between the seat of the swab mandrel and swab cup, resulting in loss of fluid. Swab cups with a rubber seat at the bottom of the cup create a good pack of seats at the point of engagement between the mandrel and the cup.

D) *Poor seated position of swab cups with the swab mandrel.* This problem may often occur due to deposits of sand grains or trash at the space between the cup and mandrel, thus preventing the swab cup from a proper seal position. Loss of fluid is due to a gap between the cup and the mandrel.

E) *Lack of evenness or smoothness of the tubing internal surface due to several factors.* Manufacturer workmanship and scale deposits are mostly causes of losing the fluid while tripping up the hole.

F) *Tubing leaks.* Depending upon the size and the location of the hole, partial or total loss of swab load is possible.

G) *Application of wrong combination swab cups and mandrels.* Using a larger size of swab cups on a small swab mandrel will cause fluid loss.

H) *Using the wrong size of swab cups within the hole.* Larger tubing size and smaller size of swab cup will result in loss of fluid.

I) *Tag fluid misjudgment.* Perhaps you did not tag your fluid at the correct depth. Therefore, sometimes you may or may not pick up a fluid load. Scattered fluid and gas pockets down the hole will create this misjudgment. Usually when you do not lift sufficient fluid load, your cable at the hoist drum will indicate that you are pulling the wire rope with no load.

J) *Upside-down swab cups run into the hole by mistake.* Upside-down cups will hardly get below the fluid. Also, these cups will not bring the swab load up the hole.

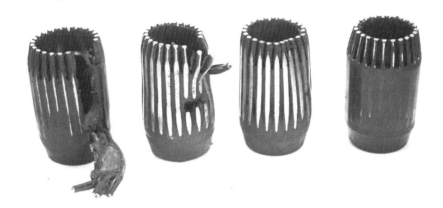

K) *Vacuum.* Sometimes a well may undergo a vacuum, which creates a dry swab run.

L) *Small and sharp arcs.* These can appear along the length of tubing string.

M) *Scattered layers of cement and scales inside the pipes.* This causes the swab cups not to seal effectively. Often this happens in a deep well that fluid cannot be brought up with rubber cups, and swab load will be lost down the hole. The swab cups will show no sign of tear; the abnormal condition of this problem usually occur inside a buckled tubing string and dirty fluid.

Selection of Swab Cups

The selection of all rubber or wire swab cups is an important judgment for the swab operator to make.

The condition of the well, the type of fluid, and the percentage of foreign material (such as sand, cement, paraffin, scale, or mud) must be noted prior to and during the swab cup selection.

Normally, if the swab operator does not have sufficient information about the wellbore condition, he may choose a normal swab cup to begin with to bring an adequate fluid load. He may choose one rubber swab cup with medium rubber lips to start with.

If the condition of the tubing and fluid subject to swab is good, the chance of selecting a wire swab cup or hard-type rubber cup may be decided.

Steel- or wire-type swab cups are built tough and strong, and they last much longer than any rubber swab cup if they are used correctly.

Some operators may worry about getting stuck by using steel swab cups. Consequently, they do not use the wire swab cups too often. This thought process may not be correct.

Wire swab cups last longer, handle more fluid, and resist abrasion and the normal wear and tear much better—longer than all the rubber-type cups.

These wire cups are bonded with medium-to-hard rubber all around, both internally and externally, and at the bottom of the swab cup seat. This is very important in sealing off fluid and leaks.

Steel swab cups are built for heavy or medium loads in deep wells. These cups are usually made up of solid-steel, metal-base steel wires; nylon; or compounded, medium-to-hard, especially formulated, high-strength rubber to provide sufficient resistance against drag, friction, and overburden from fluid load.

Wire swab cups are designed to challenge bad tubing conditions due to their flexibility, expansion, and contraction. This type of swab cup can also handle suspended sand and mud in certain quantities without problems.

One of the causes of swab cup hanging is leaking fluid, allowing the sand and scale to lodge behind the swab cups or swab bars. This problem occurs on rubber swab cups more often than wire-type swab cups. Rubber swab cups may allow foreign materials to pass, fall, and become trapped behind the cups, causing a hang-up problem. Wire swab cups do not permit fluid to pass either internally or externally due to a solid rubber base.

Even though the wire stems and the rubber covering the steel wires seem to be hard, they are very flexible under a swab load; this flexibility and expansion will create a positive seal, which wipes the pipe dry as it is coming out of the hole (tubing) and prevents any mud and floating sand from falling in the back side of the swab cups.

As far as tight spots or crooked holes, the wire cups have been tested to be superior to rubber cups. A full-sized no-go must be ran below the cup in order to clear the path for the cup going down.

During repeated runs or heavy loads, the swab cup may expand to more than the normal condition, but this expansion should not be enough to create problems. Do not run a damaged wire cup back in service. Check cups on every run to be safe.

There are two types of wire swab cups: Some swab cups are with angled or inward-bent wire stems, and some have straight wires. Use swab cups with angled or inward-bent wire stems.

Here are swab cup selections:

1) Selection of swab cup that can handle soft paraffin
2) Selection of swab cup that may handle sand (sand devil)
3) Selection of swab cup that can handle sand and scale
4) Selection of swab cup that can handle more fluid load
5) Selection of swab cup that can handle tight spots and doglegs
6) Selection of swab cup that can handle load and obstructions

The swab cups having hard rubber lips are designed to handle medium to heavy fluid loads in deeper wells. These particular cups may tear or get damaged easily if the tubing path is rough due to contact with hard-scale objects.

The swab cups with soft rubber lips are designed to handle light loads only. Soft all-rubber cups will perform satisfactorily if they are used wisely.

Accumulation of paraffin inside the tubing string forms restriction that is difficult to get down. Single soft-lip rubbers may be the choice to work with.

The soft all-rubber cups are flexible and are capable of expansion or contractions. They fit the tubing shape during the job.

Due to the elasticity of the soft rubber lips, the cups can handle minor tight spots and doglegs.

Soft all-rubber cups perform well in handling small quantities of any kind of sand, mud, cement, and paraffin without serious problems.

When to Replace a Swab Cup

This refers to all swab cups, whether a rubber or wire type.

Swab cups do not have to be broken down before replacement. Changing out a swab cup is the judgment call in part of the swab operator.

Swab cups ought to be replaced on the following occasions:

- Torn lip or lips
- Cracked or split cups (usually at the valley of the cup)
- Losing fluid, dry run
- Sign of normal wear, pull, flat, or roundness on the swab cup lips
- Exposing the swab sleeves from the bottom
- Dents, scars, or any abnormal condition that may cause the cup to pull fluid
- Changing to different types of the size of cups

The Wire Swab Cups

Wire swab cups ought to be replaced on the following occasions:

- Sufficient wire exposure due to abrasion and long work travel
- Collapsed or bent wire stems (occurs when running the cups too long with heavy loads, or vacuum force)
- Separation between the swab cups' base and the wires
- A gouge or a fluid passage between the two adjacent wire stems (caused by long, continuous trips creating a contraction after a swab cup is unloaded or cooled off)
- Change to another kind of swab cup as the job requires.

Wire Rope Flagging (Broken Wires)

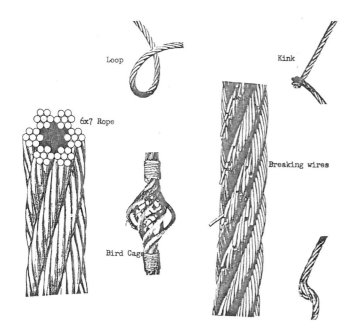

Flagging is the result of one or several broken wires due to nicks, abrasion, and improper care or misuse of the swab line. Broken wire or wires pull off the rope strand and tend to wrap around or ball up at the side of the wire rope during swab trips.

Depending on the direction of wire breakages, sometimes the broken wires ball up in the tubing, resulting in a fishing process. Often, wires ball up in the tubing lubricator and blow out the preventer and hang up. As a result, the tools will not be able to go down or pull up the hole.

Here's how to overcome some of these problems:

A) Avoid high-speed trips;

B) Always check your wire rope. Look for flat or shiny spots along the length of the line. Check and cut bad spots of the rope before putting the rope to work;

C) Do not wait until your rope pulls down or breaks apart;

D) Check the rope socket for flagging. In case of wire flag, do not wrap the rope on the drum. Slack off, force broken wires in the valleys of the rope, dry, and hold them in place with electric tape or small size wires, and then pull out carefully. After the swab tool is pulled into the lubricator, close the master valve and BOPs and lay down the lubricator with tools and cut the bad portion of your cable and resocket again; and

E) To reduce the risk of losing the cable in the hole, early detection is recommended.

Stranded Rope and Bird Nest

The major cause of bird nest may be erosion, kinks, and overstressing the swab rope. The bird nest will occur when one or more wires break apart, pack, or ball out inside the tubing and swab lubricator. Wires may separate either going down or at the time of pulling the rope out of the well. Usually, if wires separate while going down the hole and the broken wires are looking down, wires will hang the tools and will stop immediately. However, when pulling up the hole, broken wires may travel and tend to wrap around the rope in different shape and form before coming to a halt.

The bird nest will usually occur in the swab lubricator because of the pack-off assembly. If the operator sees that the swab rope is stranded, he should stop immediately, slack off, cut, or tape the wires in position. He should pull and wrap the rope on the drum. In case of a bird nest, the operator should slack off if pulling appears to be impossible.

A) Check to make sure there is no well pressure.

B) Consider any and all the safety points.

C) Leave the well open to flow line to reduce or minimize gas pressure.

D) Close the blind rams on the BOPs at the bottom section, relieve the pressure from pack-off element, and check to make sure BOPs are holding.

E) Break the top section of the lubricator off and check to see if the bird nest is above or below the BOPs.

F) Pump the pack-off tightly.

G) Pull the top section of the lubricator as high as possible, to 15–20 ft. above the wellhead.

H) Install two clamps on the sand line, just above the BOPs.

I) Using a rod elevator; set the clamps on the elevator.

J) Slack off the rope from the rope drum and lay down the swab lubricator on the ground.

K) Pull or disassemble the pack-off elements then start to pull and cut the bird nest wires inside the lubricator.

L) Completely engage the loose wires in the valley of the strand of the rope or cut and patch the cable.

M) Replace the pack-off rubber, then pick up, and install the lubricator.

N) Pump up the pack-off and install a pressure gauge. Then open the BOPs, and start out of the hole

Stuck Tools Due to Sand or Solids

Evaluation of well condition and downhole problems is of utmost importance before any attempt to swab a well. The selection of swab cups handling the sand and size of sinker bars is a critical matter. When fluid with sand is picked up by swab cups due to a fluid condition above the swab cups, sand grains tend to fall down beside the bars and on the tops of cups. Due to lip expansion, sand will not be allowed to fall behind cups unless a cut or damage occurs on the swab cups. In this case, fluid load will be divided; liquid (water or oil) will tend to fall back to the wellbore through damaged cups, gas still rising, and dry formation and/or frac sand falling on the swab cups, tend to bridge.

One of the major causes of hang-up problem in the tubing is the loss of fluid behind the swab cups. This is the result of swab cups damage, such as, fluid cut, abrasion, and wear and tear. If the cups hang up due to sand or scale, slack off, and work the tools loose, do not pull hard to squeeze the liquid off the sand grains.

Usually, when swab tools are hung up in the tubing, they are stuck at sinker bars. Large-size bars of 1¾ in. and jars are mostly responsible for hang-up problems; 1¼ in. sinker bars and smaller-size tools have sufficient clearance with adequate safety and can be used to handle sandy wells.

In gaseous wells, the chance of getting stuck is usually less. This is due to the fact that the gas tends to break the sand grains apart while pulling out of the hole and reducing swab load; however, gas without liquid will dry the sand and will make the swab job impossible. If tools are stuck with a packer in the hole, pulling the rope off the rope socket or using the rope cutter may be the final solution. Fishing a parted sand line in the tubing string and using a spear or any other fishing tools is normally a time-consuming and costly practice.

Missing the Swab Flags

As previously mentioned, the purpose of the swab flags are to caution the swab operator that swab tools are close to the surface and are expected to be in the lubricator at any moment. Avoid accidents.

It is very important to observe and react positively when the first set of flags are shown immediately at the top of the swab lubricator. Slow down and pull slowly until the last set of flags reaches the top of the lubricator. Slowly bring the tool up the hole into the wellhead and then pull inch by inch until the tool is completely cleared off the wellhead and placed inside the lubricator. Close the well immediately after the tool is cleared. Break down and check swab tools for any reason.

Missing the swab flags will certainly cause serious problems, resulting in the rope socket bumping forcefully against the oil saver plunger and bushings at the top of the lubricator and breaking the sand line or pulling the rope off the rope socket.

Severe damages to the rope can be certain. The tools will be lost in the well.

Trainees and inexperienced swab operators should be aware of this particular problem. Often sun reflection, heavy rains, and concentration of the operator's mind on other matters are the major causes of this problem, preventing the operator from seeing the flags.

Here's how to prevent this problem:

A) Stop talking to others and concentrate on the job; leave outside matters away from the job.
B) Provide three sets of flags.
C) Let someone help you if you cannot see clearly.
D) Slow down when the first flag is close to the wellhead and concentrate on arrival of swab flags.

When the operator misses the swab flags, they may cause an accident.

An accident happens so fast that the swab operator will not have a chance to react.

Tools will hit against the pack-off elements hard enough to break the cable or pull the rope off the rope socket. The cable strands will pull apart and fly in the air above the lubricator, as high as the crown blocks. The parted swab tools with some broken cable will fall back into the well with high acceleration until it will hit the fluid in the tubing string and will rest at a tight spot in the tubing or seating nipple.

The tools may be fished or pumped out of the well later.

The swab cable is called by different names in oil and gas operations: swab line, sand line, wire rope, or hoist line.

When you observe a wire rope, you may wonder how it is made. Some people may think that rope making is a simple process. Wires and rope strands are designed and engineered within a thousand-of-an-inch tolerance. Fatigue, crushing forces, ductility, brittleness, elongation, tensile strength, and other physical properties are precisely calculated by sophisticated methods.

Checking the Wire Rope by Measuring the Rope Lay

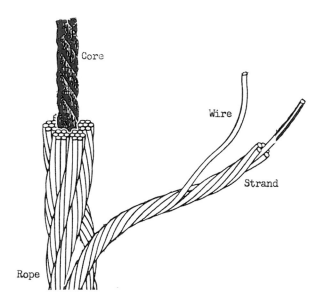

A rope lay is one complete turn of a rope strand around the wire core. Comparing the rope lay with the past rope lay traces will indicate whether the condition of the rope has changed or not. A sharp change of the rope lay without loss of rope diameter is a possible indication of loss of core or an inlaying condition.

1) Inlaying can be a result of worn sheaves or wire rope twisting.
2) Inlaying may occur due to rotation of rope. Using swivel-type rope socket will prevent the inlaying condition.

Wire Rope Break-In

New wire ropes, like any other machine, need breaking in for adjustment and conditioning of the strands for better output. Breaking in will allow the rope strands and/or individual components of your wire rope to be engaged in a set or adjusted position. Wire ropes are usually made slightly oversized. After a rope is broken in and put to service for a period of time, the increased dimension will gradually diminish due to the stretch and set position of the rope strands. The first few trips with no load or with light loads are recommended to break in the new wire rope. This process will help the rope to get in the constructional stress.

Causes of Birdcaging

The birdcaging results from a sudden shock and high stress release condition.

Sudden fluid pound released from a sudden hanging position, jarring, and so on will create this problem. Severe birdcaging will create permanent deformation of the strands, causing the strands to jump off the wire core without returning to their original position. Birdcaging spots along the length of the wire rope should be cut off before any other serious problems occur.

Physical and Chemical Characteristics of Wire Ropes

The physical characteristics of a wire rope are related to the following:

A) *Bendability and flexibility during service.* Flexibility of a wire rope is the degree a rope can be bent in an arc shape (sheave arc) without deformation. This capability of a rope depends upon the following:

   1) The larger the rope, the less flexible it is; the greater the number of wires, the more flexible is the wire rope

   2) The construction and design of the strands making up the rope.

   3) The metallic composition and chemical treatment. Galvanized ropes are much more flexible.

   4) The type of the center core. The ropes with fiber core are more flexible than the ropes with internal wire rope core (IWRC).

B) *Strength of a wire rope.* Rope strength is one of the important factors of swab rope. The strength of a rope refers to how many kilograms or pounds of tension and/or load the rope can pull without stretching or stressing. Stress load will actually transfer and divide among each wire strand in a rope and then to each single wire. Wire rope strength is usually measured in tons of 2,000 lb. or 1,000 kg of pull. Due to continuous service, a rope should not be pulled to its maximum strength value of a new rope.

Resistance against Permanent Deformation

The deformation of a rope refers to the loss of metal or wires or permanent damages due to wear, abrasion, corrosion, friction against sheaves, the bore hole, and other abnormal conditions, such as sudden fluid impact forces, shocks, and kinks. Chemical properties and cheaper steel materials will result in poor-quality wire ropes. Continuous service will cause metal loss from outside of the rope and will result in reduction of rope strength.

Metal loss from abrasion can be caused by

1) misaligned or loose sheave,
2) wrong size of sheave or pulley, or
3) improper rope spooling and drum crushing.

Wire Rope Treatment

To obtain maximum rope life with satisfactory efficiency, the swab operator should do the following:

A) *Selecting the right size and class of wire rope for the swabbing process.* There are many types and classes of manufactured wire ropes made to be used for a variety of applications. Using your wire rope in other than what the rope is recommended for could be dangerous, troublesome, and perhaps would destroy the useful life of the rope.

B) *Lubricating*

C) *Proper installation and rewinding.* Wire ropes are usually spooled by the rope manufacturer to avoid any possible damage during handling and transportation. When you order several thousand feet of cable, the manufacturer's representative will bring a new spool of wire on the location. They may rewind your new rope on your hoist drum with precision if requested.

1) Make sure the wire rope reel is set off the ground on a set of jacks with axis parallel to the ground surface.

2) The rope reel should rotate quite freely during rewinding.

3) Do not rewind the cable when the reel is stationary or sitting flat on the ground. This will simply create twisting problems.

4) During rewinding, the wire rope should be spooled tightly with evenly spaced precision to avoid any rubbing overlapping conditions when the rope is put to service.

Selection and Inspection of Sheaves

Another cause of wire rope damage could be the result of improper selection of sheave size. Crushing or damaging the wire strands may very well be the result of a smaller size of groove of the pulley sheave and is of the utmost importance. Periodic inspection of the sheaves is necessary to avoid major problems.

    A) Look for broken chipped flanges.

    B) Look for burned or flat spots in the valley of the sheave.

    C) Look for looseness of bearings on the shaft.

    D) Look for cracks.

    E) Look for an out-of-round condition of the groove.

    F) Look for scratches on the shaft and the sheave guard of flanges.

    G) Check the depth of the groove on the sheave, using wire rope groove gauges.

        1) Groove gauge for a new sheave.

        2) Groove gauge for a used sheave—this type of gauge is made to nominal diameter of the rope plus one-half the allowable rope oversize. These gauges are made to be used in the field and can be used to obtain a minimum condition of a worn groove on your sheave. When a sheave groove is worn out normally under daily operating conditions, the groove becomes deeper and usually narrow. This condition indicates that the wear is normal. This abnormal condition may be the result of misalignment of the sheave, the shaft, or broken bearings due to the wobbling of the pulley. Sometimes, if the unit is rigged up off center of a well during the swab operation, the cable will be dragged to guard and wall side, creating abnormality.

Rope Crushing

The majority of rope crushing and penning may be the result of improper, uneven wire winding on the drum. This will cause the rope wraps to scrub when partially or totally overlapping. Due to the heavy load, the rope will be forced into the space of the groove between uneven wraps, resulting in the crushing of the wires on the rope strands.

Primary, as well as secondary, spooling is very important to extend the life of your wire rope. This requires both care and skill of the swab operator. During spooling, every individual wire wrap should be tight, and individual loops should be spooled and placed correctly to avoid any possible wire damage. Since loose wraps will create kinks and twist problems, correction should be made before a wire rope deformation occurs. Notice the creaking noise during spooling, especially when you are pulling the rope swab load.

Index

Printed in the United States
By Bookmasters